FIREPOWER

JEREMY FLACK

SUNBURST BOOKS

Acknowledgements
I would like to thank my wife Julie, and my daughters, Lucy and Loretta, for all their assistance and patience in putting this book together.

Picture credits

Jeremy Flack: 6, 7t, 7br, 7bl, 9b, 10t, 11t, 12b, 18b, 21tl, 21tr, 21b, 22t, 22c, 23, 24t, 24b, 25t, 26b, 28b, 29t, 30b, 31t, 31b, 32t, 32b, 33t, 34t, 36b, 37t, 39t, 39b, 40tl, 40tr, 40b, 42tb, 42tr, 42b, 43t, 43b, 44t, 44b, 46t, 46b, 47t, 49b, 50, 51t, 57b, 58t, 59, 61, 62tl, 62tr, 62b, 64t, 67b, 71b, 72, 73t, 74

Hughes: 9t, 35t

McDonnell-Douglas: 1, 8t, 11b, 12t, 20tl, 20b, 35b, 84t, 85, 90

Hunting: 22b

Department of Defence: 18tl, 19, 20tr, 28t, 33, 78t, 78c, 78b, 79t, 79c, 79b, 80, 81, 82t, 82b, 83t, 83b, 86, 89b, 92t

General Dynamics: 18t, 38

Westland: 8b

Euromissile: 10b

LTV: 17

Aerospatiale: 13, 60, 87t, 87b, 88b

BAe: 14b, 41, 56t, 60t, 64b, 71t, 91t

Mediaphoto: 14t, 45, 76

SAAB: 15

MBB: 16t, 23

CEV: 16b

CEV/Matra Defense: 25b, 26t

Lockheed: 27, 77, 84b

Vought: 29b

Northrop: 30t

Boeing: 34b

Aermacchi: 36t

Atlas/Denel: 37b

Armscor: 47b, 56b

VSEL: 48

Royal Ordnance: 49t

MLRS Corp: 51b, 52t, 52b, 53t, 53b

Avibras: 54t, 54b, 55t, 55b

IMI· 57b

ECPA Didier Charre: 58b

Oerlikon/Contraves: 63t, 63b, 69b, 70t

Bofors: 65

Matra/CEL: 66t

Short: 66b, 67t, 68t, 68b

Alenia: 69t

John Shakespeare: 73b

Matra: 75, 89t

CFCM: 88t

Oto Melara/Matra: 92b

GFCC Marconi: 93, 94

CONTENTS

INTRODUCTION

The word 'firepower' portrays an image of great military strength. Firepower depends on the weapon system. For past generations it has mainly been dependent on guns but in more recent times the emphasis has changed to missiles, with guns now taking second place.

The destructive capability of modern firepower has accelerated at a tremendous rate. During the World War One it was measured in pounds. By World War Two this had escalated to tonnes and now we talk of kilo tonnes (KT) of explosive power from nuclear weapons.

The manpower required to operate these weapons has also decreased dramatically. The majority of the millions of soldiers that fought and died during World War One fought with machine-guns and artillery. By World War Two the firepower was provided by bombers and tanks as well as machine-guns and artillery but with reduced numbers of soldiers. During the Gulf War, the firepower was provided primarily using bombers and strike aircraft with the army making great use of rocket powered artillery requiring even lower soldier levels to produce the firepower.

During the Cold War weapons were stockpiled by the East and West to maintain a balance and 'a sense of security'. Today, with the warming of East/West relations, disarmament is proceeding and the major powers are reducing these massive stockpiles, as well as cutting defence budgets.

Disarmament is, however, hindered by the breakdown of the former Soviet Union, with the loss of control over its member states, who are claiming military equipment left there as their own. Also, terrorist or rebel groups around the world have gained access to the weapons that the major powers are getting rid of. Conflicts such as those in Azerbaidjan, Armenia, Rwanda, Somalia and Afghanistan show that there is a lot of firepower out there.

Nuclear weapons are being sold on the black market, and the question of who might use nuclear power is still present, especially in the light of such recent wars as the Gulf War in 1991 and the threatened repeat of this in 1994. Such incidences create a feeling that disarmament should be balanced – firepower needs to be retained for protection.

In this book, firepower has been divided into land, sea and air although inevitably some weapons cross over from one section into another. The aerial firepower is considered to be the most glamorous with fighter pilots able to zoom around the sky in their highly powered chargers, in one to one combat, using missiles in place of the lance. The fighter pilot is usually directed towards his target(s) by the ground or airborne radar until he is able to track the target himself. The missile will probably be launched without the other aircraft even being seen.

The bomber was the other major carrier of aerial firepower, but with the escalating costs of new technology the single role aircraft is rapidly being replaced by multi-role aircraft. The modern strike aircraft is capable of many functions, some are variants on the basic design such as the Tornado which comes in the

IDS or ground attack version or the ADV or fighter variant. Other true multi-role aircraft such as the F/A-18 Hornet as the designation suggests combine the role of fighter and attack aircraft without modification. Most strike aircraft are capable of carrying air-to-air missiles for self defence in addition to the normal strike firepower.

The helicopter now has a major place in warfare and the attack variants can be extremely effective by using night or low-light vision which can also be assisted by radar. A range of firepower can be used from guns to grenades and rockets to guided missiles.

On the ground, artillery still provides a large proportion of the firepower. However, electronics have moved in with computers to assist in the calculations required for the accurate laying of the artillery guns and tanks. In addition some ammunition have electronic sensors or guidance to improve the effectiveness of their firepower.

A relatively recent development has been the use of the rocket to replace the gun. Although artillery rockets were used in World War Two, the modern multiple launch rocket is a very sophisticated and effective producer of firepower. Using computer technology, the gunner can aim all of his rockets at different targets, fire them all within seconds of each other and immediately afterwards move out to a safe location before retaliatory firepower can be returned. Each of the rockets (and there may be ten or more) may be fitted with several hundred sub-munitions consisting of mines or small bombs. Systems such as this were deployed during the Gulf War with devastating effect but required only a small manning level.

A soldier always requires some form of firepower to protect him. This can range from the trusty rifle or machine gun through to weapon systems to deal with enemy firepower. These can include anti-tank and anti-aircraft or anti-missile weapons as well as systems to deal with minefields.

Naval capabilities are virtually infinite. Nearly all ships make heavy use of modern technology to control and operate their firepower. The ultimate source of firepower is the nuclear armed submarine, able to operate undetected in the high seas anywhere in the world and launch weapons of mass destruction at any time. These have provided the ultimate deterrent for many years and one hopes that their existence is never tested for real.

The term 'Gunboat Diplomacy' is seldom used today. The major battleships with their awesome firepower are being phased out. Today a much smaller ship can project a much more effective form of firepower in the form of missiles against ships and shore targets. The small ships are a less obvious target than the large battleships and are heavily armed with various forms of anti-aircraft weaponry and high powered guns that can even shoot down missiles. Even torpedo technology has been infiltrated by the computer to increase the effective firepower of the hunter killer submarine.

This book is a general and easy-to-use guide to some of the modern firepower of today.

AIR

TORNADO GR.1

Military attack aircraft have a major role in deploying firepower. While some attack aircraft have a specific role, many, like this Tornado are multi-role and thus capable of carrying a wide range of weapons, each of which are capable of specific functions. This RAF Tornado GR.1 weapons display includes laser-guided bombs, iron and cluster bombs, 2 in and 58 mm SNEB rockets plus their launch pods, Lepus Flare pods, Sidewinder and Skyflash air-to-air missiles as well as a range of practice munitions and their appropriate carriers. However, this is only part of the range it is capable of carrying. Most current multi-role aircraft have an internally mounted gun which can be used on ground targets or for close range air-to-air combat. The Tornado is fitted with a pair of the Mauser 27 mm cannon.

PHANTOM FG.1

Other aircraft which do not have an integral gun may have one fitted for specific roles as required. Some versions of the McDonnell-Douglas F-4 Phantom have no gun fitted. Instead this RAF Phantom FG.1 has a General Electric 20 mm Vulcan cannon mounted in a belly-mounted pod. This cannon is a six-barrel gatling type gun which has the capability of firing 6,000 rounds per minute.

The Phantom is another multi-role aircraft and the Vulcan cannon is only one of the weapon systems that it is capable of carrying. This aircraft can carry other weapons, including air-to-air missiles bombs and rockets, as well as having a reconnaissance role.

The Canadians produce a 70 mm diameter CRV-7 while the French have a range of 68 mm and 100 mm rockets. Italy manufactures up to 122 mm while Spain goes down to 37 mm. As well as variations in size and performance, their warhead can also vary according to the task. In addition to those with an explosive charge, some rockets contain a number of sub-munitions in the form of metal darts which are capable of armour piercing.

These rockets can be fired as a salvo where a number of the rockets are required to hit a concentrated target. This is being demonstrated by an RAF Harrier.

SNEB ROCKETS

The air-to-ground rocket evolved during World War Two and was little more than a small rocket-powered bomb. A large number of these rockets have been developed over the years by many countries.

Here, Thomson-Brandt 68 mm SNEB rockets are being loaded into a Matra launcher pod prior to a training mission. These rockets have been widely used by various NATO forces.

70 MM ROCKETS
Alternatively rockets can be ripple launched. The US Army AH-64A Apache (right) is firing the heavier 70 mm rocket while the Blackhawk (below) fires 2.75 in rockets. These rockets are unguided as a rule and would be used against 'soft' targets.

HUGHES AGM-65 MAVERICK

For the armoured and specific 'hard' targets a number of guided missiles have been designed. The TV-guided missile has a TV camera fitted into the nose and transmits the picture back to the launch aircraft. The picture is watched by the weapon systems officer who uses a small joy-stick to fly the missile onto the target.

The Hughes AGM-65 Maverick is typical of this type of missile and is seen here being launched from a USAF McDonnell-Douglas F-15E Strike Eagle. Often these missiles are developed into a family and included in the Maverick range are infra-red and laser-guided missiles.

Right: The Maverick was used in substantial numbers during the Gulf War against the large number of Iraqi armoured vehicles. This Russian built 2S1 of the Iraqi Army illustrates the destructive power of these missiles.

HUGHES TOW

The use of the helicopter for attack missions was developed during the Vietnam War when missiles and guns were fitted and proved to be effective against enemy positions. A large number of anti-tank missiles have been developed for helicopter operations of which some started life as ground-launched missiles.

The Hughes TOW (Tube-operated, Optically tracked, Wire-guided) is one such Anti-Tank Guided Weapon (ATGW). Once launched, the missile is guided onto the target by the gunner/weapon systems operator who views the target through a sight. The command signals are transmitted to the missile from a small joy-stick via a fine command wire making it resistant to all forms of jamming. This British Army Air Corps Lynx is being re-armed with TOW missiles during a replen (replenishment of fuel and armaments) during operations in Kuwait. The Lynx can be fitted with up to eight of these missiles.

The TOW ATGW can also be fitted to vehicles.

AEROSPATIALE HOT

The Aerospatiale HOT is another ATGW developed from a ground-launched system. Seen during the launch from a French Army Gazelle, the HOT also trails a command wire and the HOT 2 variant is capable of piercing up to 1,200 mm of solid steel.

MCDONNELL-DOUGLAS AH-64A APACHE

The primary attack helicopter of the US Army is the McDonnell-Douglas AH-64A Apache. This helicopter is capable of projecting a formidable firepower against any enemy land forces. It is capable of carrying up to 16 Rockwell AGM-114 Hellfire tactical air-to-surface missiles on stub wings although a mixed load can be carried. This Apache is fitted with eight Hellfire plus two rocket pods. Using laser designation, Hellfire is a fire-and-forget missile system which is capable of locating its target even when the target is not initially visible.

AH-64D LONGBOW APACHE

The AH-64D Longbow Apache is fitted with the distinctive mast mounted Longbow Fire Control Radar for better weapon delivery. It is seen here launching a Hellfire missile. GEC-Marconi have developed a variation on Hellfire names. Brimstone is fitted with a millimetric wave seeker which is virtually undetectable to the target. Brimstone can be fitted to Harriers as well as helicopters and vehicles.

BELL 0H-58D KIOWA

The Bell OH-58D Kiowa was originally bought by the US army as a scout helicopter to call up the heavily armed Apache once suitable targets were located. The latest OH-58D Kiowa Warrior can be fitted with a mast-mounted sight giving mechanically stabilised day and night vision capabilities using optical, TV and Thermal Imaging as well as a laser range finder/designator. This OH-58D Kiowa Warrior is launching a pair of Hellfire missiles that can be fitted to the Kiowa. In addition, a pair of 2.75 in rocket pods can be fitted or alternatively a machine gun pod or a pair of Stinger air-to-air missiles.

Bofors have developed the RBS-17 as a coastal defence missile from Hellfire in conjunction with Rockwell.

MIL MI.24 'HIND D'

Similar systems have been developed by the Russians. This Czech Air Force Mil Mi.24 'Hind D' is fitted with 57 mm rocket pods as well as rails for four of the 9M17P ATGW referred to as the 'AT-2 Swatter' by NATO. It is also fitted with a 9-A-624 12.7 mm four-barrel gatling type gun which is capable of 4,000 to 5,000 rounds per minute. In addition to these weapons the Mi.24 inventory of weapons includes the 9M114 – 'AT-6 Spiral' ATGW, as well as a range of gun and grenade pods plus bombs from 50 kg up to 500 kg and cluster bombs.

AEROSPATIALE AS.15TT

The Aerospatiale AS.15TT (Tous Temps or all weather) is an anti-ship missile which, like the Penguin has been developed for use by ship and aircraft. Similarly, a coastal defence variant has also been proposed. The AS.15TT relies on guidance data provided by the launch aircraft radar such as the Agrion fitted to this Aerospatiale AS.365F Dauphin 2.

A development of the AS15 is the MM15 which is intended to be fitted to fast patrol boats and other small craft as an anti-ship missile.

KONGSBERG PENGUIN MK.2 ANTI-SHIP MISSILE

The Norwegian Kongsberg anti-ship missile was originally developed as a ship launched weapon for use by small, fast naval craft. The Penguin Mk.2 MOD 7 was specially adapted for use with the US Navy Sikorsky SH-60B Skyhawk (above) as the AGM-119B. Further variants of the Penguin were developed for the Royal Norwegian Air Force and carried by F-16s as the Mk.3 plus another proposed for coastal defence. Penguin is a fire-and-forget missile which can be programmed with a target together with deceptive way points. When fired, it will find the target using its infra-red (IR) seeker. It is claimed to be resistant to decoys.

BAe ALARM

The British Aerospace ALARM is an Air Launched Anti-Radar Missile designed to help provide the RAF with a deep penetration capability. Developed and built in conjunction with Marconi, ALARM was produced as a British alternative to the US High-speed Anti-Radar Missile. It is virtually an autonomous system which is launched when the threat receiver warns of enemy radar activity or that targets might exist.

ALARM has several modes of operation including direct attack at the radar source. Another mode directs the missile to an altitude where a parachute deploys. The missile seeker then waits for a suitable target to be identified, jettisons the parachute and descends onto the target. This mode would allow the aircraft to transit the area with a greatly reduced risk of being detected and attacked by SAMs (surface-to-air missiles).

SAAB RB-15F

The SAAB RB-15F is a long range anti-ship missile weighing about 600 kg. It is equipped with an advanced guidance system which enables it to accurately approach the target area where the search and acquisition system can locate the target. It is powered by turbo-jet and has a range of approximately 150 km. The RB-15 is operational with the Royal Swedish Air Force with the AJ-37 Viggen and a pair can be seen here fitted to the JAS.39 Gripen in addition to the Maverick and Sidewinder missiles.

TEXAS HARM AGM-88

The US High-speed Anti-Radar Missile AGM-88 HARM was designed to replace the AGM-45 Shrike which was used extensively during the Vietnam War and on a number of occasions since. As with the Shrike, the USAF fly the HARM on specially designated squadrons of Phantoms, referred to as the 'Wild Weasel'. These aircraft either operate on their own, locating and destroying enemy radar installations or will fly in conjunction with a force of strike aircraft to provide protection especially from radars associated with SAMs.

The HARM is used by the USAF on the F-4G Phantom and the US Navy on F/A-18 Hornets and EA-6B Prowlers and was used in considerable numbers during the Gulf War. In addition the German Air Force has purchased a quantity which are fitted onto the Tornado.

DASSAULT MIRAGE IV

Aircraft have been used to carry bombs since the early days but have been limited in their destructive power by the sheer weight of the weapons. The advent of the atomic and then nuclear weapons changed this and all the nuclear powers have deployed at least part of their nuclear weaponry on aircraft.

This French Air Force Dassault Mirage IV, which is using a rocket assisted take off, has an ASMP (Air-Sol-Moyenne-Portee) medium range air-to-surface tactical nuclear missile fitted under its fuselage. Capable of a 300 km range, this missile has a Mach 2 to 3 performance and carries a warhead of some 300 KT. The Mirage IV is being augmented by the Mirage 2000N and will eventually be replaced by it. The ASMP can also be carried by the French Navy Super Etendard.

Development of the ASMP as a conventional or an anti-ship missile has been proposed.

LTV A-7 CORSAIR II

While nuclear weaponry can provide the ultimate in destructive power, the iron bomb still remains a vital part of any military armoury. They retain the flexibility of being carried by a manned aircraft which can select its target and then release.

These bombs are being dropped by US Navy A-7s and were used to assist the Allied war effort during Desert Storm when the A-7s dropped a total of 1033 tonnes of iron bombs on Iraqi positions.

GENERAL DYNAMICS F-111

Above: For some attack profiles the aircraft will be required to fly at low level to avoid detection by enemy radars. These bombs are therefore fitted with a brake chute to ensure that the aircraft – in this case an F-111 of the **USAF** – will be clear before the bombs hit the target. It also ensures that the bombs do not ricochet off the target as could well happen with a conventional iron bomb.

Left: The General Dynamics F-111 is in fact a bomber and not a fighter as indicated by its designation. Developed to be the first operational swing-wing bomber, the F-111 had a turbulent evolution with much political wrangling, followed by the loss of half of a deployment of six aircraft to Vietnam in only four weeks. Following modifications a further deployment of 48 F-111s to Vietnam in 1972–73 saw the aircraft fly over 4,000 combat missions with only six losses.

The F-111 proved to be a capable aircraft with the strike on Libya and numerous missions conducted during Desert Storm when 94 were deployed. The F-111 is capable of carrying up to 50 x 825 lb bombs.

IRON BOMB
Loading a 1,000 lb iron bomb onto an **RAF Tornado GR.1**.

MCDONNELL-DOUGLAS F-4 PHANTOM

The F-4 Phantom has been a highly successful product of the McDonnell-Douglas stable. This multi-role aircraft is capable of carrying a total of 16,000 lb (7,257 kg) of ordnance is seen here dropping 18 x 825 lb bombs. It was widely used during the Vietnam war as a bomber and a fighter as well as its Wild Weasel activities.

MCDONNELL-DOUGLAS AV-8B HARRIER
These retarded bombs are being dropped by a **USMC AV-8B Harrier** and are fitted with folding fins which deploy to provide the braking force for low level operations.

MCDONNELL-DOUGLAS F/A-18 HORNET
The iron bomb is a relatively cheap weapon to produce an explosive force at the target. In good hands, with a good weapons delivery system, it can be surprisingly accurate.

The F/A-18 Hornets of **VFA-82** are being armed with iron bombs aboard **USS** America prior to another mission against Iraqi forces during Desert Storm.

MCDONNELL-DOUGLAS F/A-18D HORNET
An F/A-18D of the USMC VMFA(AW)-533 drops a 2,000 lb bomb during training over a Nevada range.

BOEING B-52 STRATOFORTRESS

Iron bombs on the wing pylons of a B-52 awaiting the next mission against Iraqi forces. The B-52, originally designed in the late 1940s as a long range strategic bomber, entered service in 1955 and has remained in front line service ever since, despite the construction of the prototypes of two replacement aircraft – the XB-70 and the B-1A.

BOEING B-52 STRATOFORTRESS

Despite its age the B-52 remains capable of delivering a staggering 70, 000 lb (31,800 kg) of ordnance. While it has a large bomb bay, the B-52 can also be fitted with under-wing pylons on which further bombs can be carried.

BOEING B-52 STRATOFORTRESS

Loading bombs onto a B-52 during Desert Storm. These aircraft could carry 51 x 750 lb M117 bombs. Operating in formations of three aircraft, these B-52s cleared areas on the ground including minefields, 1.6 km wide and 2.4 km long. It was estimated that the B-52s were responsible for 31 per cent of the 62,000 tonnes of bombs that were dropped during the conflict.

BOEING B-52 STRATOFORTRESS
As in most conflicts, the bombs acquired a quantity of graffiti to help them on their way.

BOEING B-52 STRATOFORTRESS
A total of eight B-52Gs were deployed to RAF Fairford from various Wings to make up the 806th BMW (Provisional). They flew a total of 60 combat sorties in 19 days during which 1,158 tonnes of munitions were dropped. Each flight lasted an average of 16.3 hours and required a total of 6 million kg of fuel which was transferred from KC-135 tankers during the deployment.

HUNTING BL755
The cluster bomb is a relatively new concept. The munition consists of a bomb-shaped container which splits open at a set altitude and scatters a quantity of sub-munitions. The principal of all cluster bombs is similar although the nature of the sub-munitions can vary according to the mission role. Some may be designed for hard targets while others for soft or a combination. Also sub-submunitions can have sensors which guide them to their targets.

The Hunting BL-755 contains 147 separate sub-munitions which, once deployed, are parachute retarded. They detonate on impact and this fires a high velocity slug capable of penetrating steel as well as creating fragmentations from the case.

MBB MW-1

The **MBB MW-1** is in effect a 'super' cluster bomb and can deploy up to 4,536 projectiles. The MW-1 comprises a large pod which is fitted under the fuselage of a Tornado. It consists of four sections of 28 transverse tubes into which the sub-munitions are loaded.

The aircraft is flown low over the target, and charges of different strengths ensure a uniform distribution of the sub-munitions. A single pass can cover an area from 55 to 500 m wide and 200 to 2,500 m long. A range of sub-munitions can be fitted including armour-piercing, anti-tank mines, fragmentation, area denial and runway cratering. The aircrew can adjust the weapon spread and density prior to firing.

A development of this system has been designed to fit standard bomb racks of which a variant is being developed for the Royal Swedish Air Force for the SAAB Gripen. Several similar dispenser weapon systems have been designed including the French Alkan Models 500 and 530, Brunswick LAD and Hunting JP233. Further advanced designs include the Franco/German Apache, a stand off dispenser weapon which obviates the requirement of the aircraft to overfly the target, thus resulting in a safer mission profile. The RAF lost a number of their Tornados while delivering JP233 during Desert Storm.

LASER-GUIDED BOMB

The advent of the laser has brought about a tremendous advance in weapon delivery. These **RAF** airmen manhandle an inert Laser-Guided Bomb (LGB) during a loading exercise on a Harrier GR.7.

The **LGB** comprises three parts – the nose containing the laser guidance and control system, the bomb, and the bolt-on tail section.

TEXAS PAVEWAY LGB

The **RAF** system is a development of the Texas Paveway system used by the **USAF**. The system works by a designator illuminating the target with a coded laser light. This can be from the same or another aircraft or even from a ground source. The aircraft will launch the powerless bomb into what is, in effect, a cone of entry. The guidance sensor will locate the reflected light from the target and guide the bomb to the apex of the cone using the small winglets.

TEXAS PAVEWAY LGB

A typical load for the **F-111F** – four LGBs together with a pair of **AIM-9** Sidewinder air-to-air missiles for self protection. Under the fuselage can be seen the **Pave Tack** equipment used for target acquisition and illumination.

HUGHES GBU

The Hughes **GBU** series of weapons is similar in principal to the Laser Guided Bomb (**LGB**), but instead of requiring the laser designator the seeker head includes either a **TV** camera for daylight attacks or the **Infra-Red** seeker as fitted to the **AGM-65** Maverick for night attacks.

This **GBU-15** is fitted to a 2,000 lb bomb on a F-111F. On the 17 January 1991, the first day of Desert Storm, 53 F-111Fs were launched against Iraqi targets using a range of **GBU** and **LGB** weapons.

FRENCH MATRA BGL

The French Matra **BGLs** (Bombes a Guidage Laser) being dropped here by French Air Force Jaguars are similar in design and operation to the Texas Paveway system.

MATRA BGL
A 1-tonne **Matra BGL** released from a Mirage 2000 is considered ideal for the demolition of bridges and piers in order to prevent an enemy advance. A 400 kg **BGL** has been developed and can also be used in the anti-ship role.

MATRA DURANDEL

The Matra Durandel has been designed for concrete penetration to disable airfields by destroying runways and taxiways. This weapon is dropped at high speed, over 1000 km/hr; and at low level, 100 m, over the target. It immediately deploys a parachute to attain its vertical attack mode, so that there is no chance of a ricochet and then the rocket powered missile accelerates to a high velocity enabling it to penetrate concrete over 40 cm thick before exploding. The resulting heave of the concrete slab renders that part of the runway or taxiway very difficult to repair quickly unlike a crater which only requires filling. Some warheads can be fitted with a timed delay ensuring that the repairs cannot be commenced immediately.

The Durandel is used by many air forces around the world as well as the French Air Force. These Durandels are fitted to a **USAF F-111F**.

LOCKHEED F-117A

The Lockheed F-117A is a USAF aircraft. It is actually a stealth bomber and not a fighter as inferred by the designation. It is illustrated here being prepared for a bombing mission. This highly secret aircraft remained elusive to the world during its first five years of operational service – until November 1988. During the build up to the Gulf War some 40 of the 59 built were deployed to the region. When the war began, these stealthy aircraft were able to fly undetected to targets. During the first 24 hours of Desert Storm the F-117s were reputed to have hit 31 per cent of all targets although they accounted for only 2.5 per cent of the aircraft used.

The F-117 is subsonic – but relies on its unique design of a number of angled flat surfaces together with a radar-wave absorbing finish to remain undetected by enemy radars. The Saudis referred to it as 'Shabba' or ghost.

The advantage of using the F-117 is that it reduces the number of aircraft required to complete a given mission. A standard mission with iron bombs might require 32 attack aircraft. To protect these would require 16 fighters plus eight Wild Weasels and four radar jammers. This whole force would then require 15 tankers to provide fuel for them all.

The same result could be achieved using just eight F-117s plus two tankers!

F-117
An F-117 dropping a 2,000 lb Laser-Guided Bomb. During the Desert Storm similar weapons were used with pin-point accuracy to penetrate 12 m of concrete – previously thought impregnable to conventional bombs. The result of accurate bombing on a huge HAS (Hardened Aircraft Shelter) at the Kuwaiti airfield of Al Salem during its occupation by the Iraqis.

ROCKWELL B-1B LANCER

The Rockwell **B-1B** Lancer entered service with the **USAF** as a multi-role, long-range strategic bomber in 1985. The Lancer can carry a wide range of bombs and missiles including up to 20 air-launched cruise missiles.

This **B-1B** Lancer is being prepared for a mission at RAF Fairford during its first deployment to Europe. Seen here, about to be loaded, are inert practice bombs made from concrete.

NORTHROP B-2A

The **Northrop B-2A** is the latest **USAF** strategic penetration bomber and like the **F-117** utilises stealth technology.

NORTHROP B-2A
The B-2A drops a single 2,000 lb bomb over a test range. The B-2A is capable of carrying various weapon loads
including up to 36,590 kg bombs for the conventional role, 16 **AGM-69 SRAM II**, or strategic/tactical nuclear weapons.

A-10A THUNDERBOLT II
The **A-10A Thunderbolt II** is a
dedicated tank buster. It has an
integrally mounted 30 mm Avenger
gatling gun in the nose. This is
capable of firing 3,900 rounds per
minute and can include depleted
uranium shells which are capable of
defeating most armour. It has 11
pylons onto which a total of 6,640
kg of weapons can be fitted.

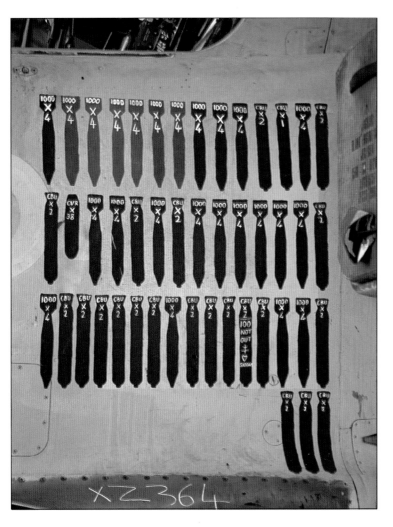

SEPECAT JAGUAR GR.1A

Allied strike aircraft were kept very busy throughout Desert Storm. The RAF Jaguars which entered service in 1973 are beginning to show their age. However, they were tasked with a total of 158 missions involving 617 sorties and over 921 combat flying hours. The bulk of the ammunition expended were the 1,000 lb bombs (total 750) and the CBU-87 Rockeye cluster bomb (total 385). In addition eight BL-755s were dropped and 32 pods of the CRV-770 mm rockets were fired (total 608 rockets). At the cessation of hostilities all the aircraft returned home to RAF Coltishall, with only a little minor battle damage.

In time honoured tradition this Jaguar has its mission tally painted on the nose. On each of the 47 missions, symbols are marked according to the type of munitions dropped. These include the majority with the 1,000 lb iron bombs, a few with the CBU-87 Rockeye cluster bombs and one sortie with two CRV-7 rocket pods loaded with 19 rockets.

PANAVIA TORNADO GR.1

Mission marking on an RAF Tornado GR.1 following an operation in Desert Storm. These symbols indicate the number of JP233, iron bomb and Laser Guided Bomb missions that this particular aircraft flew operationally.

SEPECAT JAGUAR GR.1A

A substantial force of **NATO** aircraft are based around the Adriatic area, either on carriers or airfields. Their task is to ensure an economic and military blockade of the former Yugoslavian states and to ensure the safety of the members of **UNPROFOR**. To this end various fighters and strike aircraft patrol the region.

This **RAF** Jaguar is armed with bombs and missiles while it flies over Bosnia. This acts as a reminder to the various warlords of the ability of the **UNPROFOR** Commanders to call immediate air attacks if required and hopefully acts as a deterrent against aggressive actions.

BAe SEA HARRIER FRS.1

Likewise, a Royal Navy Sea Harrier from **HMS Invincible** or **Ark Royal** is tasked with the same role. During such deployments, **HMS Ark Royal** was nick-named the Martini carrier due to its ability to position in clear weather and provide aircraft on call at 'any time, any place, anywhere'.

LOCKHEED AC-130 SPECTRE

Although built as a transport aircraft, the Lockheed C-130 Hercules has fulfilled numerous functions over the years. The most aggressive is the AC-130 Spectre gunship, designed to replace the AC-47 gunships used in Vietnam.

The AC-130H Spectre is fitted with a range of infra-red and low light TV sensors. They are armed with two 20 mm Vulcan gatling type guns, a 40 mm cannon plus a 105 mm recoiless gun. When called forward they can lay an impressive amount of firepower onto a target while circling it. The cockpit is fitted with a side looking Head-Up Display (HUD) to assist the pilots during night missions.

Photographed when refuelling from a USAF KC-135 Stratotanker over the Adriatic, this AC-130 is part of the NATO force on patrol over Bosnia. The various guns can be seen protruding from the fuselage.

LOCKHEED AC-130 SPECTRE
The AC-130 Spectre fires its Vulcan gun at a ground target.

USAF BOEING B-52 AND TUPOLEV TU-95 BEAR

Old adversaries meet. The end of the Cold War and the warming of East/West relations has seen numerous events that would not have seemed possible only a few years previously. Here a USAF Boeing B-52 is parked on the same dispersal as a Russian Tupolev Tu-95 code named 'Bear' by NATO.

Both of these aircraft were designed in the late 1940s to early 1950s as long range strategic bombers and have remained in service ever since, despite the comings and goings of various replacements. Interestingly the Bear's ancestry can be traced directly to the Tu-20 – a reverse engineered copy of the B-29 – which was designed and built by Boeing.

BOEING AGM-86 ALCM

The Boeing AGM-86 Air Launched Cruise Missile (ALCM) is a long range missile with a sophisticated navigation system enabling it to fly a confusing non-direct route to a target. This navigation system can verify its position by comparing a radar picture of the surrounding landscape with a stored computer model and make any necessary adjustments. It is then able to accurately hit its designated target.

The ALCM is fitted with a nuclear warhead and 20 ALCMs can be carried by a single B-52 or B-1B.

HUGHES AMRAAM

Above and below: The Advanced Medium Range Air-to-Air Missile (AMRAAM) shown here is being launched from a McDonnell-Douglas F/A-18 Hornet. It has been designed to replace the AIM-7 Sparrow in service with the US armed forces. The AMRAAM is designed to have a performance which is far in advance of the Sparrow, yet lighter and cheaper. The guidance system has state-of-the-art technology with active and passive seekers to ensure that it has a high hit probability, without the launch aircraft becoming vulnerable by providing the illumination radar.

AIM-9 SIDEWINDER

The **AIM-9 S**idewinder is probably the most produced short range air-to-air missile in the world considering the number of variants and copies that have been built. The latest variant of this missile has shown significant advances in its capabilities. It is able to hit a target from any direction, somewhat different to the original missile developed in the 1950s which could only be fired from the rear of a target and could be distracted by the sun or reflections from a lake.

IRAQI HELICOPTER

The remains of an Iraqi helicopter destroyed during the Gulf War. This helicopter was one of many aircraft destroyed by the Allied forces and shows the destructive power of air-to-air weapons.

WESTLAND LYNX AH.7

The British Army Westland Lynx AH.7 that were deployed to the Gulf for Desert Storm were fitted with pintle mounted **GPMG** machine guns to give extra defensive or offensive firepower should it be required.

ATLAS ROOIVALK

The Atlas Rooivalk is a South African built attack helicopter. Like the Apache and Cobra, it can carry anti-tank missiles and is fitted with a chin-mounted cannon. It can also operate the Kukri air-to-air missile. This missile contains a conventional infra-red sensor but its tracking system is electronically linked to the pilot's helmet. This enables the pilot to simply look at the target and fire the missile once the missile confirms that it is tracking the target.

LAND

GENERAL DYNAMICS M1A1 ABRAMS

The tank emerged onto the battle field in 1915 during World War One. It was armed with machine guns and a small gun. The main battle tank (MBT) of today is a highly technical machine. About the only thing remaining in common with the original designs are the tracked means of mobility.

The modern MBT is constructed of state-of-the art armour to provide protection for the crew. It is fitted with sophisticated communications, fire control and navigation systems. It has infra-red surveillance capability for night and poor visibility and a laser range finder for an accurate first round hit.

The General Dynamics M1A1 Abrams entered service in 1979 and since then over 8000 of these MBTs have been built for the US Army. 1956 M1A1s were deployed to the Gulf for Desert Storm. Four were disabled, four were damaged but repairable but none were destroyed by enemy action. Although seven crews reported taking direct hits from Iraqi T72s with 125 mm guns, these Abrams suffered no serious damage.

The M1A1 (illustrated) is currently being augmented by the improved M1A2 and it is planned that a large number of the M1A1s will be upgraded to a similar standard.

VICKERS CHALLENGER 1

The British Army is equipped with over 400 Challenger 1s which have traditionally been based in Germany. This original version of Challenger is fitted with a 120 mm gun which will eventually be replaced with the new gun of the Challenger 2.

During Desert Storm, 150 Challenger 1s of the British army were deployed. Fitted with reactive and passive armour, these MBTs proved to be highly effective and were responsible for the destruction of some 300 Iraqi MBTs without a single loss.

VICKERS CHIEFTAIN

The Chieftain has been the MBT of the British Army since 1967 due to its powerful capabilities which have been improved over the years through various modification programmes. In addition to those built for the British Army, the Chieftain has served with the armies of Iran, Iraq, Jordan, Kuwait and Oman.

Kuwait took delivery of 143 Chieftain MBTs. These were in action against the Iraqi forces during the summer of 1990 but were overwhelmed by the superior numbers of the invaders.

VICKERS CHALLENGER 2

The Vickers Challenger 2 is an upgrade of the Challenger MBT which has been in service with the British Army since the early 1980s. Challenger 2 is a substantial improvement over the existing Challenger I which was highly effective during the Gulf War.

Challenger 2 has a new turret, incorporating second generation Chobham armour and a new gun barrel. This barrel is chrome-lined to give it an extended life and maintain consistent accuracy.

The Challenger 2 is also fitted with a range of sensors and sights to ensure that it remains fully capable in all battle conditions. These various systems are linked via the latest generation fire control computer enabling accurate first shot hits to be achieved.

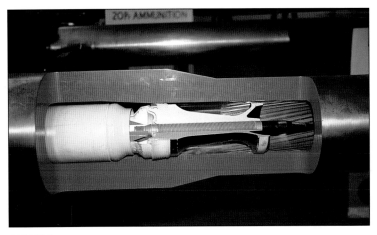

APFSDS

The Armour-Piercing Fin-Stabilised Discarding Sabot (APFSDS) is a lethal weapon. It is made from tungsten and is extremely heavy. This projectile travels like a dart at its target and uses its kinetic energy to penetrate any armour.

120 MM AMMUNITION

British 120 mm ammunition is supplied in two parts with the propellant separate from the projectile. The propellant is supplied in bags which are completely combustible when ignited. The lower weight is easier to manhandle, there is no cartridge case to dispose of, the risk of fumes is greatly reduced and storage space is reduced.

The projectiles come in various forms. From left to right is the APFSDS with tungsten dart protruding from the sabot. The black and yellow shell is the HESH (High Explosive Squash Head) which effectively sticks to armour and gives the warhead time to burn through the armour. The green shell produces an opaque smoke barrier. The blue shell is PRAC SH – a practice form of HESH. On the far right is the DS/T PRAC – a practice form of the APFSDS.

M60 MBT

The M60 MBT was originally ordered for production in 1959 and entered service with the US Army in 1960. It is fitted with a 105 mm gun and can carry up to 63 rounds of ammunition. When production was completed, more than 15,000 M60s had been built.

A considerable number of M60s were deployed to the Gulf where they participated in Desert Storm. These were operated by the Saudi Arabian and Egyptian armies as well as the USMC.

This line up of M60s, ARV and M115s show just a tip of the iceberg of the vast quantities of armour and other vehicles that were deployed to the region. The amount of cargo transported was more than four times that for the D-Day operations.

LONG TOM
Some older equipment simply soldiers on. This American-built 155 mm 'Long Tom' was captured from the Iraqis during the Gulf War and the manufacturing stamp was dated 1945.

VSEL FH-70
FH-70 was designed to meet the requirements of German and US armies for a 155 mm howitzer to replace their existing 155 mm M114 and the British Army 140 mm guns. Capable of fitting into a C-130 Hercules or carried by a CH-47 Chinook, the FH-70 has a range of over 24 km with standard ammunition or over 31 km with base-bleed.

155 MM AMMUNITION

Like the 120 mm, British 155 mm ammunition is supplied in two parts: the propellant and the projectile. This propellant is supplied in bags as a full charge and when loaded enables maximum range to be achieved. If the bag is opened, a series of smaller coloured bags are accessed. By removing the appropriate coloured bag a reduced range can be achieved.

Various projectiles are available to suit different tasks. These include High Explosive (HE) which has a devastating fragmentation effect, Base Ejection Smoke which discharges four smoke canisters and gives a dense cloud of smoke, and Illuminating which can provide 2.1 million candelas to illuminate the enemy at night. The base-bleed ammunition is fitted with a small solid fuel rocket that extends its range. Illustrated is a sectioned M77 shell showing the sub-munitions.

2S1 SELF-PROPELLED HOWITZER

The former Soviet Union designed and built 2S1 is a 122 mm Self-Propelled Howitzer. It entered service with the Soviet and Polish armies in 1971. Although not confirmed, it is thought that some 10,000 2S1s have been built and supplied worldwide with nearly 8,000 being declared by the Russians as part of the Conventional Forces Europe Treaty.

The size of a complete round of ammunition can be seen on this captured Iraqi Army 2S1. Various types of ammunition can be fired by the 2S1 of which the HE has a maximum range of 15.3 km.

RDM M114/39

The howitzer provides a means of delivering firepower using a high trajectory. Centuries ago this weapon would be used to attack the inside of walled cities. Today, it is widely used to provide indirect firepower (the target not visible to the gunner) and can cause major damage to target areas. The M114/39 is a modification of the existing M114 howitzer using a 39 calibre barrel by RDM of the Netherlands in place of the existing shorter 23 calibre barrel. This lengthening of the barrel gives a small increase in range with the same 155 mm ammunition. But the effect is really apparent with the use of extended range ammunition when maximum range can be doubled.

VSEL AS90

Chestnut Troop of the 1st Regiment Royal Horse Artillery at its war establishment strength. Eight AS90s provide the main firepower and the rest of the vehicles provide the back up. This includes the Battery Commander and his team, a reconnaissance party, 12 DROPS ammunition vehicles and assorted support vehicles.

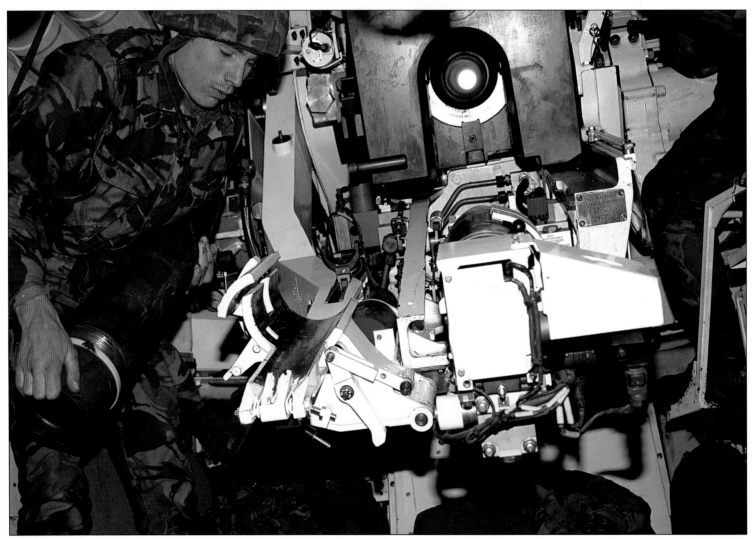

VSEL AS90

The AS90 carries 48 rounds of ammunition. A battery of AS90 is capable of delivering 24 x 44 kg rounds or over 1 tonne of ammunitio onto a target in under 10 seconds. It has a range of 24 km with standard ammunition or 30 km with extended range munitions.

There are plans to fit a longer barrel onto AS90 in the future, which will give a 40 km range. Other developments include the integration of the turret onto other tank chassis via a simple adapter ring, putting new life into what would be considered obsolete tanks.

VSEL AS90

**Although it still requires some brute strength, the AS90 is fitted with an advanced loading
and laying system which enables a burst rate of three rounds in under 10 seconds.**

ARMSCOR G5

**The Armscor G5 is a 155 mm towed gun howitzer which uses extended range projectiles
to give it a maximum range of 39 km. It is capable of a 30 km range with standard ammunition
and a maximum firing rate of three rounds per minute can be achieved over 15 minutes.**

VSEL 155 MM UFH

The VSEL 155 mm Ultralightweight Field Howitzer (UFH) can provide the firepower of a conventional howitzer. However, through a revolutionary design and the use of lightweight materials, a strong and stable weapon platform has been achieved at less than 4 tonnes.

This results in the ease of transport of the UFH, typically by a 2.5 tonne four-wheel drive vehicle, tactical air transport or medium lift helicopter. This now makes the UFH available to Rapid Deployment Forces increasing their effective firepower.

ROYAL ORDNANCE LIGHT TOWED HOWITZER

The Royal Ordnance 155 mm Light Towed Howitzer (LTH) is also designed to meet the weight requirements for helicopter deployment by the Rapid Deployment Forces; with the 155 mm artillery capability to potentially replace the 105 mm Light Gun. The LTH has a maximum range of 24 km which can be increased to 30 km using extended range ammunition. It has a similar weight to the UFH.

ROYAL ORDNANCE LIGHT GUN

The 105 mm Light Gun achieved the satisfactory compromise of firepower and mobility and achieved a substantial number of orders from various armies including those of Australia, Botswana, Brazil, Brunei, Ireland, Kenya, Malawi, Morocco, New Zealand, Oman, United Arab Emirates, UK and USA.

The Light Gun is capable of a burst fire rate of eight rounds per minute although the sustained rate is normally three. This gun was used extensively during the Falklands conflict when transversing very soft ground would have created a potentially major obstacle for heavier weapons.

ROYAL ORDNANCE LIGHT GUN
The Royal Ordnance 105 mm Light Gun has a range of between 2.3 km and 17.2 km. This gun can be effectively used in the direct (target visible) and indirect role. The large difference in range is achieved by a combination of barrel elevation and the Charge Bag system.

LTV MGM-52 LANCE

The rocket did not become a serious threat until late in World War Two when the Germans started to use the V-1 and V-2s. However, these had no precise guidance systems and were merely a nuisance to the military. The same could be said about the Iraqi **SCUD** missiles that were fired during Desert Storm.

During the 1950s and 1960s improvements were made to the guidance systems enabling a fair degree of accuracy. A product of early 1960s design was the **LTV MGM-52 Lance** which did not enter service with the **US Army** until 1971. Capable of carrying a conventional or nuclear warhead, Lance had a range of between 5 and 120 km. Guidance details are fed into the missile before launch which is either from a self-propelled launcher based on the M113 chassis or a lightweight towed trailer. The missile is fired and rapidly reaches Mach 3, then the rocket motor cuts and the rest of the flight is unpowered.

LTV MLRS

Typical of a modern development of the rocket as a means of delivering a warhead is the **LTV Multiple Launch Rocket System (MLRS)** – a cross between the conventional rocket and artillery. The MLRS is a highly mobile weapon system, capable of launching any quantity of its load of 12 rockets at separate specific targets in quick succession. It can then move to another location before the enemy has time to react and return fire, often referred to as 'shoot and scoot'.

LTV MLRS

The MLRS has proved to be a popular system, currently operational in the armies of Bahrain, France, Germany, Italy, Netherlands, Turkey and US with further sales to Greece and Japan. In 1992, just 10 years after the first rocket was completed, the 500,000th rocket was delivered to the US Army.

LTV MLRS

The first operational deployment of MLRS to Desert Storm attracted great interest. Each rocket is able to spread its sub-munitions over an area the size of a football pitch. As a result, the MLRS rockets were known as black rain by the Iraqi troops.

LTV ATACMS

MLRS is used to fire the **US Army TAC**tical **Missile System (ATACMS)**. With a range in excess of 100 km, initially this missile is used to deliver 950 sub-munitions. Later sub-munitions will be terminally guided. It substantially extends the range of existing **MLRS** rockets and other cannons, rockets or artillery missiles and thus is considered to raise the nuclear threshold. **ATACMS** is being used to replace Lance and was fired operationally during Desert Storm.

LTV MLRS

A single launcher with a crew of three, in firing its complement of 12 rockets, could cover between 12 to 24 hectares with nearly 8,000 grenade like sub-munitions. One Iraqi commander was quoted as saying that he saw all but seven of his 78 guns put out of action by these bomblets. A total of 246 launchers were deployed to the Gulf by the US and British Army.

The new AT2 rocket is fitted with 28 anti-tank mines which are capable of penetrating all known tank hulls. The mines can be programmed with six different operating times for self-destruct if not activated, creating a highly hazardous task of clearance. Two MLRS launchers can lay a minefield 1.5 km by 400 m at a range of 35 km with 336 mines in just one minute.

Another development is for an extended range rocket for MLRS. The first of these has already been fired at a range of over 45 km. Future developments being considered include terminally guided sub-munitions.

AVIBRAS ASTROS II
The Brazilian Avibras **ASTROS II** is an Artillery **Sa**Turation **RO**cket **S**ystem. Using the
127 mm–SS30, 180 mm–SS40 and 300 mm–SS60, **ASTROS II** has a range of 10 to 80 km.

AVIBRAS ASTROS II
The **ASTROS II** was used by the Iraqi Army during the Iraq/Iran war and again during
the Desert Storm. Saudi Arabia used her **ASTROS II** to assist in the re-taking of Khafji.

AVIBRAS ASTROS ROCKET
During Desert Storm large quantities of the **ASTROS** rockets were launched. Depending on the rocket used and the firing conditions, between 1.25 to 10.5 sq km of soft and hard targets can be destroyed or neutralised by a single 16 second attack.

AVIBRAS ASTROS
The **ASTROS SS40** and **SS60** rockets can be fitted with sub-munition warheads which are capable of perforating 220 mm of steel plate. As such the **ASTROS II** becomes an effective coastal defence system being able to attack approaching warships or against disembarking troops during an amphibious assault.

RAYO MULTIPLE ROCKET SYSTEM
The 160 mm Rayo Multiple Rocket System is a joint programme between the Chilean **FAMAE** and Royal Ordnance to produce a weapon that retains the firepower capability of conventional rocket systems without the complexities or cost for the Chilean Army.

ARMSCOR VALKIRI
The South African Armscor Valkiri short range artillery rocket system consists of a 6 x 6 chassis with a mine-proof cab for crew protection. It is fitted with a 40 tube rocket pack, from which the rockets can be fired singly or in ripples. The 127 mm rockets have a range of 36 km and can be fitted with various warheads including anti-personnel with some 8,000 steel balls and can cover 1,000 sq m with lethal shrapnel, anti-armour and sub-munitions.

A 12 rocket trailer variant is under development with a maximum range of 5.5 km while naval use has also been proposed.

IMI LAR 160

The **LAR 160** Light Artillery Rocket system built by Israel Military Industries entered service in 1984. It utilises an **AMX-13** chassis into which a launcher has been fitted which can accept two launch pod containers each of which contains 19 rockets. To re-arm, the empty container is removed and a fresh one fitted taking less than 10 minutes. Each rocket can be fitted with various 50 kg warheads including sub-munitions. All 36 rockets can be fired within 60 seconds and have a range of 30 km.

Variations on this system include other chassis with similar or reduced rocket capability. An eight-rocket trailer is also under development while another proposal is to fit 25 rocket containers aboard warships.

BM-21b

The 122 mm **BM-21a** is a Soviet designed and constructed rocket system which entered service in 1964 and has been widely exported.

This captured Iraqi **BM-21b** has a 36 rocket capacity and takes just 18 seconds to fire all its rockets. The warheads fitted contain HE-fragmentation, incendiary or smoke, or can be fitted with a range of chemical agents.

9M28 ROCKETS

These dumped Iraqi 122 mm 9M28 rockets have a range of 10.8 km and take two crew approximately five minutes to reload the BM-21.

The Soviet Spesnatz and other specialist units have been equipped with single missile launchers.

AEROSPATIALE MILAN

Once launched the Milan is guided onto the target by the gunner keeping the cross hairs in the view finder on the target. Flight control commands are transmitted down a fine wire which is trailed from the missile. The warhead contains a solid charge giving it a good penetration capability. This widely used weapon has been significantly improved over the years. Initially French designed, the Milan project was joined by Germany and subsequently built under licence in the UK. It is operational in many countries throughout the world.

The Milan warhead can be seen here emerging from the launchers during Desert Storm. This specially shaped warhead is designed to activate the reactive armour before the actual warhead strikes. Milan has been recorded as having penetrated 106 mm of steel armour.

The Milan launcher can also be fitted to various vehicles.

AEROSPATIALE MILAN

The anti-tank weapon saw rapid development during World War Two as a means for the infantry to combat the tanks which had by then become a major threat.

As with most weapons systems, the tank versus anti-tank weapon is a constant leap frog of technology. As the means of destroying tanks evolve, so the tank designers improve the tank design accordingly. The modern MBT and armoured vehicles have a high degree of defensive armour and those used during Desert Storm by the Allies were 'up-armoured' with additional bolt-on layers of armour. In addition several systems have been built which consist of small bolt-on devices referred to as 'reactive armour'. When hit by a projectile the whole device explodes creating a counter force which effectively causes the projectile to be blown off resulting in next to no damage.

The Aerospatiale Milan is an Anti-Tank Guided Weapon (ATGW) which has been in production since 1972. This man portable system comprises a tripod-mounted firing post which incorporates an infra-red tracker plus an optical sight and launching ramp with a firing grip.

BAe SWINGFIRE

The BAe Swingfire is a long range ATGW with a wire guidance control. This form of guidance control ensures that the missile cannot be jammed. Although it is possible to mount Swingfire on various vehicles the British Army examples are all fitted to Alvis Striker.

The launch box's missiles are loaded into bins and can be fired from within the vehicle or from a vantage point up to 100 m away from the concealed vehicle.

Swingfire has a range of up to 4 km and is capable of destroying the armour on any known tank.

AEROSPATIALE ERYX

The short range Aerospatiale ERYX ATGW is designed for easy use by infantrymen. With a low-velocity launch, the missile can be safely fired from the shoulder and from within an enclosed space. Guidance is via a control wire and simply requires the sight to be kept on the target for up to four seconds which is all the time the missile will take to travel its 600 m range.

LAW

The British Light Anti-armour Weapon (LAW) is a single shot disposable weapon for the infantryman. By simply removing the two end caps and extending the tube the weapon is ready to fire.

Accuracy is a major factor for the designers of LAW and it has been fitted with a five round spotting rifle. With a maximum range of 500 m, LAW has a high hit probability.

AT-3 SAGGER

These captured Iraqi anti-tank weapons at a disposal dump include various projectiles for the RPG-7 and SPG-9 anti-tank launchers and a couple of AT-3 'Sagger' ATGW. The AT-3 is a wire-guided missile usually carried by BRDM-2 or BMP-1/2 vehicles or 'Hind' helicopters. It is thought to have a range of between 500 m and 3 km.

PG-7

A British Army Officer holds a PG-7 round from a RPG-7 launcher. This was photographed in an Iraqi weapons store captured in Kuwait.

The PG-7 is a Soviet built short-range anti-tank weapon which entered service in 1962. It has been built in considerable numbers and widely used in the former Warsaw Pact countries, Asia and Africa.

SEPECAT JAGUAR

The effectiveness of ground firepower against aircraft can have considerable effect on the outcome of a conflict. This French Air Force Jaguar was photographed in Saudi Arabia shortly after the ending of the Gulf War. It has been hit by a surface-to-air missile which has destroyed one of the engines. Fortunately for the pilot his second engine continued to function and he was able to land the aircraft safely. Further close inspection of the cockpit canopy showed evidence of another encounter with the enemy. This time from small arms fire where a bullet had been shot diagonally through the cockpit and very narrowly missed the pilot's head.

OERLIKON/CONTRAVES SKYGUARD

Guns have been used to try to hit aircraft since they first appeared on the battlefield. Initially rifles and machine guns were used but anti-aircraft firepower was not developed until World War Two when there was a need to combat the large numbers of enemy bombers.

The Swiss Oerlikon 35 mm air defence gun has been reliable and effective. It has remained in production for many years with deliveries being made to many countries.

The current package includes the addition of a Contraves Skyguard fire control system which is capable of automatically acquiring and tracking low flying targets, conducting an IFF (Identify Friend or Foe) check, and engaging an enemy target with 550 x 35 mm rounds per minute from each of its two barrels.

OERLIKON/CONTRAVES SKYGUARD

The Oerlikon/Contraves Skyguard air defence system uses various types of 35 mm high explosive shells. This illustration shows the effect of the conventional nose mounted fuse. An alternative is the base-mounted fuse which delays the explosion momentarily to cause maximum internal and structural damage.

While a large number of the Oerlikon 35 mm guns were built many years ago they still remain an effective weapon and Oerlikon/Contraves are offering an overhauling and modernising service for these guns to give them about another 20 years of life.

BAe RAPIER

The guns have increased their firepower over the years and the ammunition rounds are a relatively inexpensive projectile. The Germans were the first to experiment with anti-aircraft rockets but today the guided missiles have a much higher rate of success.

The BAe Rapier began as a private venture with the old British Aircraft Corporation (BAC) in the early 1960s. It first entered service with the British Army in 1971 and the RAF in 1974. Over the years the detection and tracking systems for Rapier have been updated. However the missile, which is often referred to as a hittile, has required little improvement.

This is because the missile has proved to be extremely accurate. It contains a very small amount of explosive (0.55 kg) as it relies on its ability to actually hit the target. The kinetic energy from a 42.5 kg missile hitting an aircraft at Mach 2 creates considerable damage.

BAe LASERFIRE

BAe Laserfire is a version of the Rapier surface-to-air missile system. The unique millimetric surveillance radar cannot be detected or jammed by any current or forseeable countermeasures equipment. It is therefore invisible to the enemy and cannot be attacked by an anti-radiation missile.

Once a target is detected by the radar, the laser automatically acquires and tracks the target. The inboard computer calculates the firing solution and gives the operator a 'free to fire' indication. The operator confirms that the target is hostile and the missile is launched. The dual action and fragmentation warhead enable a high probability kill while the new active laser proximity fuse can engage small **RPV** types of targets.

To date, some 400 Rapier systems have been delivered to the Armed Forces of Abu Dhabi, Australia, Brunei, Indonesia, Iran, Oman, Qatar, Singapore, Switzerland, UK, USA and Turkey.

BOFORS RBS.70
The Swedish Bofors RBS.70 is a guided air defence missile using a laser beam for guidance control. It is capable of hitting very low flying helicopters as well as aircraft. It has no electromagnetic radiation, making the fire unit hard to detect.

MATRA MISTRAL

The Matra Mistral is a light-weight anti-aircraft missile which was designed from the outset to be operated from land, sea or air. The various systems enable the Mistral missile to be launched from a variety of platforms – Santal is a six round armoured turret while Guardian is a joint project with Boeing to fit Mistral to the HMMWV (High-Mobility, Multi-purpose Wheeled Vehicle). The Sadral and Simbad are both naval applications for Mistral. ATAM (Air To Air Mistral) was developed for helicopter operations for use against enemy helicopters and even combat aircraft. A rush programme was conducted to enable the Mistral to be fitted onto the French Gazelle being deployed on Operation Daguet (the French element of Desert Storm) where they flew escort and air-to-air combat missions.

The MANPADS tripod launched, portable, single-ammunition system is illustrated while Alamo and Atlas enable the Mistral to be vehicle mounted. Currently over 10,000 missiles are operational in twelve different countries.

SHORT STARBURST

A cut away of the Short Starburst in its sealed storage/launch container. The shape on the missile is a giveaway to its Javelin ancestry.

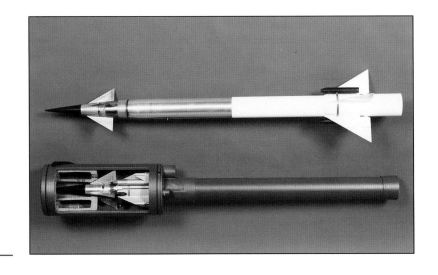

SHORT STARBURST

Short Starburst is a laser-guided, all-weather, anti-aircraft missile and was developed from the highly successful Javelin missile. Due to its laser guidance, Starburst is immune to all known forms of countermeasure. Starburst can be shoulder launched with the sealed missile container clipped onto the aiming unit.

Alternatively a Lightweight Multiple Launcher (LML) provides a multi-engagement capability with three missiles mounted on a launcher post. A further variant of the LML can be vehicle mounted.

Such is the accuracy of the Javelin and Starburst missiles that during training the radio controlled target frequently gets hit although it is a fraction of the size of a real aircraft.

SHORT STARSTREAK

The Short Starstreak is a high velocity missile which contains three very accurate and highly manoeuvrable darts. A two stage motor accelerates this lethal salvo towards its target. Within a fraction of a second the missile has reached its maximum speed of several times the speed of sound and the darts separate to home in on the target under the control of the aimer.

Shorts have teamed with Boeing to fit Starstreak onto the Avenger for the US Armed Forces.

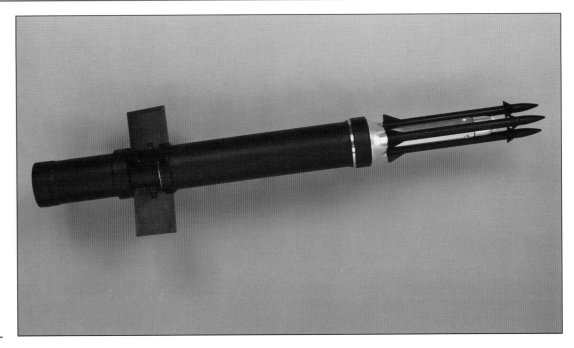

SHORT STARSTREAK

Removed from its storage and launch container, this view of the Starstreak clearly shows the three darts which contain a highly explosive warhead.

ALENIA ASPIDE

The multi-role Italian Alenia Aspide surface-to-air missile for use against low and very low flying aircraft. Developed from the AIM -7 Sparrow, Aspide has been developed to operate from the Albatros naval point defence system and the Spada ground based point defence system (illustrated). An air-to-air version is being developed for the Eurofighter 2000 as an alternative to the AMRAAM missile.

OERLIKON/CONTRAVE AHEAD
The Oerlikon/Contrave AHEAD is an advanced 35 mm ammunition which fires a number of heavy metal sub-projectiles as it approaches its target.

OERLIKON/CONTRAVES AHEAD
The effect of the sub-projectiles from the AHEAD round hitting a Falcon missile target is quite devastating and should have destroyed the missiles guidance system and fusing.

ADATS

The Air Defence Anti-Tank System (ADATS) is an international project for a single missile that can be used in the anti-aircraft and anti-tank role. The system has been mounted onto a M113 chassis although it is capable of being fitted to various vehicles. The system is equipped with a search radar, forward looking infra-red and TV sensors and uses a laser beam for missile control. The missile has a dual-purpose warhead with a proximity fuse for the anti-aircraft role and impact fuse for ground targets.

During live firing evaluations ADATS has proved to have excellent reliability and lethality. During US Army Forward Area Air Defence System (FAADS) evaluations ADATS achieved eight out of ten hits while its nearest competitor only managed five.

BAe MERLIN

The principle of the mortar dated back many centuries and consists of a high angled gun much like a howitzer to fire relatively short distances. Over the years, technology has enabled this to be a more predictable weapon and the modern mortar can be surprisingly accurate and, with the range of bombs available, quite versatile.

Modern technology has increased the effectiveness still further.

The BAe Merlin is close to becoming a missile. This terminally guided mortar munition has a small electronic processor which operates a millimetric radar as well as handling the safety and arming. Stabilising fins are extended while the radar searches for moving armoured vehicles, failing this it scans for stationary ones. Once detected the processor steers the mortar using small canards on the nose.

MACHINE GUN

Despite all the modern technologies available to the military planners, the infantryman and his gun still plays a vital part of any army. The 7.62 mm GPMG machine gun can provide a reasonable level of firepower with a rate of fire of 625-750 rounds per minute.

GIANT VIPER

Although the Giant Viper is specifically designed to clear a path for tanks, it can be used to breech any minefield. Giant Viper consists of a trailer with a launch rail, a projector and a 229 m hose packed with plastic explosive. When required for use, the trailer is towed to within 45 m of the edge of the suspected minefield. The projector is loaded with eight rockets.

GIANT VIPER
The projector is fired across the minefield and pulls the hose with it.

GIANT VIPER
Once the hose has dropped onto the ground the hose is detonated and the result is a huge blast which will clear a path 7.28 m wide and 189 m long. If this is of insufficient length a further Giant Viper can be towed up the path and the procedure repeated.

MBT (MAIN BATTLE TANK)
While the weight of ammunition is a problem for the infantryman the same cannot be said for the MBT.
The Chieftain is fitted with two machine guns and normally 6,000 rounds of 7.62 mm ammunition would be carried.

OTO MELARA/MATRA
A development of the OTOMAT has been built to satisfy a coastal defence requirement.

SEA

KONGSBERG PENGUIN MK2 MOD 7

The Norwegian Kongsberg anti-ship missile was originally developed as a ship launched weapon for use by small, fast naval craft. The Penguin Mk.2 MOD 7 was specially adapted for use with the US Navy Sikorsky SH-60B Skyhawk (see page 14) as the AGM-119B.

Further variants of the Penguin were developed for the Royal Norwegian Air Force and carried by F-16s as the Mk.3 plus another proposed for coastal defence.

Penguin is a fire-and-forget missile which can be programmed with a target together with deceptive way points. When fired, it will find the target using its infra-red (IR) seeker. It is claimed to be resistant to decoys.

LOCKHEED TRIDENT

The Trident is the successor to the Polaris strategic missiles and was first tested in 1987. As with Polaris, Trident can be launched from a submerged submarine and can carry multiple nuclear warheads over a range of some 7,000 km.

Following the East/West reconciliation, consideration is being given to changing these warheads from nuclear to conventional. The use of nuclear arms cannot be justified against a non-nuclear power, and therefore cannot be used as a deterrent. However, a conventionally armed missile is acceptable.

MCDONNELL-DOUGLAS BGM-109C

The capability of the Tomahawk is illustrated in the following sequence. This **BGM-109C** had been launched from a submarine 640 km out to sea. Utilising its integral **TER**rain **CO**ntour Matching capability (**TERCOM**), it compared the flight path with the programmed route and made any necessary corrections during its flight.

The accuracy of the hit against the structure together with the destructive power of the 1,000 lb warhead make this an extremely effective weapon system. Latest models are also capable of achieving a time-on-target to enable a surprise coordinated attack, use **GPS** for greater accuracy and have an extended range. The **BGM-109D** carried 166 sub-munitions.

BATTLESHIP

The ultimate of visual firepower of recent years has to be the US Navy battleships firing their 16 in main batteries.

The first of the USS Iowa Class of battleships was commissioned in 1943 with a further three entering service over the following sixteen months. Each were armed with nine 16 in guns plus twenty 5 in guns and all took part in the Pacific Campaign of World War Two. Even at that time they were unique – capable of 33 knots – they were the fastest battleships ever built. They were heavily armoured with a 12 in main

belt plus thick deck armour and numerous water-tight compartments that were to make sure that they could survive a number of direct hits. At the end of the war they were de-activated and put into storage.

In 1950 they were re-activated to participate in the Korean war but were subsequently de-activated again.

Here they are seen firing the Numbers 1, 2 and 3 Mark 7 16 in guns during a main battery firing exercise aboard the battleship USS Iowa.

BATTLESHIP

When the former Soviet Navy's Kirov class battleship appeared on the high seas these great ships were re-activated once more from 1982. Before doing so these battleships underwent a refit which entailed modernising the electrics, accommodation and some weaponry.

During the refit, four of the 5 in guns were removed and replaced with launchers for the Tomahawk and Harpoon missiles and Phalanx guns were added for self defence. Unfortunately USS Iowa suffered an explosion in a turret which caused major loss of life and resulted in her final de-activation along with that of USS New Jersey.

BATTLESHIP

This left just **USS Missouri** and **Wisconsin** in service. When the Iraqi invasion of Kuwait took place these two warships joined the large number of ships that were deployed to the Middle East. When Desert Storm commenced, **USS Missouri** (illustrated) and **USS Wisconsin** were both used to launch Tomahawk cruise missiles against Iraqi targets.

When the war was over these two ships were de-activated as part of the post-Cold War dearmament process.

VICKERS MK8 NAVAL GUN

The largest gun now fitted to Royal Navy operational warships is the Vickers 4.5 in Mk.8 Naval Gun. This is currently operational with ten classes of warships in six different countries. It is a fully automatic, remotely controlled system capable of firing 25 rounds per minute, each one weighing 21 kg. It is capable of devastating bombardment because of its high level of accuracy, high fire rate and long range.

Left: LOCKHEED POLARIS MISSILE

Both the US and former Soviet Union have fielded large numbers of nuclear armed land- and sea-based missile systems. Missiles have always acted as very effective deterrents in warfare.

The submarine has been developed to carry rockets since the 1950s. Gradually they have become more powerful in terms of rocket power and in warhead power. The Polaris missile has now been retired from US Naval use and replaced by the Trident.

However, the Polaris missile still remains in service with the Royal Navy Resolution Class of nuclear powered submarine, although it is currently being replaced by the new Vanguard Class, which is armed with Tridents.

Below: MCDONNELL-DOUGLAS AGM-84 HARPOON

The development of missile guidance has enabled the design of sophisticated weapons that can fly independently of any aircraft guidance. Many can hit a ship without being distracted by countermeasures.

The McDonnell-Douglas AGM-84 Harpoon is typical of this class of weapon. Initially designed to be an air-launched missile, the capability of launching either from ship or submarine was subsequently added. This turbo-jet powered missile has a range in excess of 67 nautical miles and is fitted with a 221.6 kg warhead.

MCDONNELL-DOUGLAS HARPOON

The target information is fed into the guidance system prior to launch. Once launched with the aid of a solid fuel booster, the Harpoon settles down to a sea skimming height. As it approaches the target the missile locks onto the target and initiates a snap up manoeuvre followed by a swoop down onto the target.

Harpoon is in service with the **USAF, USN, USMC** and the Royal Navy.

MCDONNELL-DOUGLAS HARPOON
**The Destroyer USS Arthur W. Radford launches a Harpoon missile in the Gulf
during exercises just a few days before the commencement of Desert Storm.**

AEROSPATIALE SM.39 EXOCET

The **Exocet** anti-ship missile commenced production in the early 1970s. This Aerospatiale sea-skimming missile has been developed as a family with various launch modes but with a common mode of attack. With the exception of the air-launched version, each missile is stored and ready to fire in a sealed launch container.

The **SM.39 Exocet** is a submarine-launched missile which is fired from a standard torpedo tube. An underwater vehicle propels and guides the missile towards the target. The whole weapon emerges from the water at 45 degrees, the underwater vessel separates and the Exocet missile conducts its conventional attack in which it uses an active radar homing head for terminal guidance.

AEROSPATIALE MM.40 EXOCET

The **MM.40 Exocet** is the surface-to-surface member of the family. Seen here being launched from a French Navy Frigate, the **MM.40** is capable of being fired from most classes of ship and is operated by over 25 navies throughout the world.

AEROSPATIALE MM.40 EXOCET

A close up of the MM.40 Exocet as it emerges from the sealed launch container. In addition to land- and sea-launched versions, the **AM.39** Exocet is air-launched by helicopter, combat or maritime patrol aircraft. It was this variant which achieved world attention when Argentine Navy Super Etendards launched the missiles against the Royal Navy Task Force off the Falklands, sinking the destroyer **HMS** Sheffield and the freighter Atlantic Conveyor.

Below: AEROSPATIALE MM.40 BLOCK 2

Further development of the Exocet has resulted in the availability of the enhanced MM.40 Block 2 as a coastal defence battery.

OTO MELARA/MATRA OTOMAT MK.3

The Oto Melara/Matra **OTOMAT** Mk.3 is a turbo-jet powered anti-ship missile with a low infra-red signature and a maximum range of 180 km. Capable of speeds up to Mach 0.9, the **OTOMAT** can have up to three way-points fed into its navigation system and can deliver its 205 kg warhead from a sea-skimming or pop-up-and-dive attack.

The **OTOMAT** is designed to be used to complement the **MILAS** torpedo carrying missile system.

MCDONNELL-DOUGLAS BGM-109B TOMAHAWK

The launch of the **BGM-109B** Tomahawk from the nuclear-powered attack submarine **USS La Jolla** of the **US Navy**. This anti-ship variant of the Tomahawk has a range of approximately 480 km.

MCDONNELL-DOUGLAS BGM-109B TOMAHAWK

The launch of a McDonnell-Douglas Tomahawk cruise missile to a target that might be anything up to 2,400 km away.

Cruise missiles can be launched from land, sea or air, some from all three. The navigation system is pre-programmed with the accurate coordinates of the target and also gives a number of way points. Once launched, the missile will fly at a subsonic speed to each of the way-points, checking its route with a computerised terrain map.

Tomahawk is probably the best known cruise missile. It is a stand off, deep strike weapon capable of being launched from land-based launchers, ships or submarines against land and sea targets. During Desert Storm some 300 Tomahawks were launched at Iraqi targets.

On TV news reports, during the Gulf War, missiles could be seen flying down streets to meet their target.

Four basic variants of the Tomahawk have been built. The first was the BGM-109A which was designed to deliver a nuclear weapon against a land target. BGM-109B delivers conventional weapons for the anti-shipping role, BGM-109C delivers conventional weapons against a land target and the BGM-109D delivers sub-munitions against a land target.

The BGM-109B, once launched, will fly a path which tries to conceal both its launch point and its ultimate target. It utilises a similar guidance system to the Harpoon with its passive identification and direction finding equipment capable of seeking out, locking on and striking targets over a 480 km range with a 1,000 lb warhead.

BAe VERTICAL LAUNCH SEAWOLF

BAe Vertical Launch Seawolf is a ship launched anti-aircraft and anti-missile missile. The vertical launching of the missile eliminates any blind spots and increases the rate of fire that can be sustained against multiple synchronised attacks. This missile was developed from the conventionally launched Seawolf missile already in service with the Royal Navy and is currently fitted to all the Type 23 frigates

The Vertical Launch Seawolf is stored in silos. When activated the missile rises vertically clear of the ship and is then rotated to a horizontal flight path by the boost motor and thrust vector control which are jettisoned shortly after launch. A solid fuel motor gives rapid acceleration to over Mach 2 to intercept and destroy the target.

SIGNAAL GOALKEEPER CIWS

The Martin Marietta/Signaal Goalkeeper anti-ship missile defence is one of a number of Close In Weapons Systems (CIWS). Similar systems include the Selinia-Elsag Dardo, Thomson-CSF Satan, GE/SAGEM SAMOS, IAI TCM-30, BMRC Seaguard, and General Dynamics Phalanx, most of which are based on the gatling type gun.

The Goalkeeper can be considered to be one of the more effective systems with a radar controlled GAU-8/A gun which can fire about 4,200 rounds of 30 mm missile piercing, discarding sabot ammunition per minute.

TORPEDO

The development of the torpedo has been similar to that of the anti-tank weapons although over a much longer period. The first torpedoes were developed about 100 years ago and were very primitive. They required a great deal of luck to hit the target.

OTO MALERA/MATRA MILAS

The Oto Malera/Matra **MILAS** anti-submarine missile is currently under development for the French and Italian navies. Designed to complement the **OTOMAT** anti-ship missile, **MILAS** is a torpedo carrying missile with a **50 km** range. It is designed to be fired from the same launching tube as the **OTOMAT** enabling any combination of the missiles to be loaded on board ship. A single control console is used to fire either missile while the single combat system takes input from the sonar or surface target detectors. This shows the firing of a development **MILAS**.

GEC MARCONI TIGERFISH

The **GEC Marconi Tigerfish** seen being loaded into a Royal Navy attack submarine. The Tigerfish is a heavy weight torpedo which is wire guided as well as having acoustic homing. It is fitted with a computer which can receive information direct from the fire control system aboard the submarine. As the torpedo approaches the target the active sonar should detect the target and make any corrections.

The effect of a Tigerfish on HMS Lowestoft, an obsolete Royal Navy frigate being used for target practice.

INDEX